NORTHBOROUGH PRIMARY SCHOOL
CHURCH STREET
NORTHBOROUGH
PETERBOROUGH PE9 9BN
01733 252204

Slopes

written by Caroline Rush
and
illustrated by Mike Gordon

Wayland

Simple Technology

Wheels and Cogs
Slopes

Levers
Pulleys

Series Editor: Catherine Baxter
Book Editors: Sue Barraclough and Sarah Doughty

First published in 1996 by
Wayland (Publishers) Ltd
61 Western Road, Hove
East Sussex, BN3 1JD, England

British Library Cataloguing in Publication Data
Rush, Caroline
Slopes. – (Simple technology)
1. Slopes (Inclined planes) – Juvenile literature
I. Title II. Gordon, Mike, 1948 –
      551.4'36

ISBN 0 7502 1865 7

Typeset by MacGuru
Printed and bound in Italy by G. Canale and C.S.p.A., Turin, Italy

Contents

If you had to climb a mountain, how would you do it?

Would you climb straight up a steep mountainside?

Or take a longer route following a gentle slope?

Both ways would take you to the top of the mountain, but which would take less effort?

Slopes can make moving upwards
a lot easier. The Egyptians built slopes
to help them build the pyramids.

It was a lot easier to move the
huge blocks of stone up a slope.
They could not lift them from
the ground.

8

We use slopes in our homes to travel from one floor to another. Stairs are slopes made from steps.

Ramps are slopes.
They can be used to raise people...

... and vehicles from one level to another.

11

12

Coming down a slope can be fun!
A helter-skelter is a long slope that goes
round and round.

Some tools use slopes.

A screw has a long sloping edge, called a
thread, that winds around in a spiral.
It's a bit like a helter-skelter!

Cut out a paper triangle.
The long side is the slope.

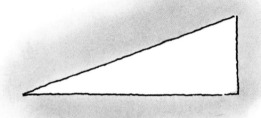

Wind it around a pencil. Can you see how
the slope winds around in a spiral shape,
like the thread on a screw?

Try pushing a nail into a piece of wood.
It is not possible without a hammer.

But you can use a screwdriver to push
a screw into a piece of wood.

When you turn the screw, the thread travels a long way. It goes around and around in the wood, and the screw moves in a little way. But it takes much less effort than pushing something straight in.

You use screws all the time.
When you turn a tap you
are turning a hidden screw.
This opens and closes the
hole that lets water through.

A corkscrew is a tool for getting a tightly fitting cork out of a bottle.

A wedge is another useful type of slope.
It is useful because its shape means it can be
pushed between two surfaces.

A wooden wedge
can be used to make
something stop
moving.

An axe is a metal wedge. Its shape makes splitting an object in half easy. When the thin end enters the object its shape pushes the two sides apart.

In the same way that slopes help you to move upwards easily, they can help you to travel downwards quickly.

Put a toy car on a flat surface.
Now raise one end of the surface.
What happens to the car?

Now try this experiment.

You will need:
Building blocks
wooden plank
toy car
tape measure
chalk

First of all rest one end of the plank on one
block. Put the car at the top of the slope.
Notice how fast the car travels. Then
measure the distance that the car travelled.

Add more blocks and watch the car roll down the slope again. What do you notice about the speed of the car and the distance travelled as the slope is increased?

Make a penny drop game.

You will need
balsa wood (1 cm and 0.5 cm thick)
PVA glue
ruler
pencil
balsa wood knife

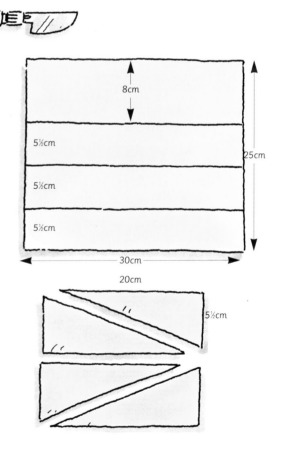

1 Ask a grown-up to help you cut out a baseboard from the balsa wood 0.5 cm thick. Mark lines across the board with a pencil.

2 Cut out 4 triangles from balsa wood, 1 cm thick.

3. Stick them on to the lines on the baseboard with the short edges butting up to the sides. Leave to dry.

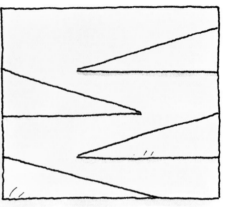

4. Cut out 4 more strips of balsa wood, 0.5 cm thick. Two should be 25 cm x 5 cm and the other two 30 cm x 5 cm.

5. Stick the strips around the edge of the baseboard to make a frame.

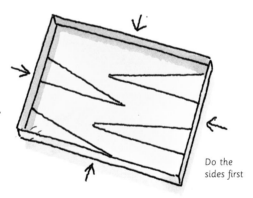

Do the sides first

Paint it brightly. Then tilt your penny drop game to make a penny run down the slopes.

Glossary

Effort   The force used to do work, eg lifting a heavy object.

Pyramid   A large triangular building which the ancient Egyptians built to hold the body of a dead king or queen.

Ramp   A slope used as a machine to make work easier, also called an inclined plane.

Screw   A nail with a spiral groove around it so that it can be put in a hole and twisted, used to fasten things together.

Notes for adults

Simple Technology is a series of elementary books designed to introduce young children to the everyday machines that make our lives easier, and the basic principles behind them.

For millions of years people have been inventing and using machines to make work easier. These machines have been constantly modified and redesigned over the years to make them more sophisticated and more successful at their task. This is really what technology is all about. It is the process of applying knowledge to make work easier.

In these books, children are encouraged to explore the early inspirations for machines, and the process of modification that has brought them forward in their current state, and in doing so, come to an understanding of the design process.

The simple text and humorous illustrations give a clear explanation of how these machines actually work, and experiments and activities give suggestions for further practical exploration.

Suggestions for further activities

- Make a collection of slopes either in actual or picture form and discuss their uses. Include less obvious slopes such as screws and nuts and bolts.

- Investigate the principle of friction. Test out slopes with different surfaces and record and display your results.

- Explore the principle of gravity to encourage an understanding of why coming down a slope is easier than going up a slope.

- Visit a playground and go down a slide. How can the children slow down or speed up a journey? Roll a ball down a slide. Does this travel faster than a person? Ask the children why they think this is.

Books to read

Machines at Work by Alan Ward (Watts, 1993)
How things Work by Brian Knapp (Atlantic Europe Publishing, 1991)
Experiment with Movement by Brian Murphy (Watts, 1991)
Simple Science/Push and Pull by Mike and Maria Gordon (Wayland, 1995)

Adult reference
The Way things Work by David Macaulay (Dorling Kindersley, 1988)
Available as CD Rom (Dorling Kindersley, 1994)

Index